A BUG BOOK for

_____

Lady bug, lady bug
fly away home.

Nursery Rhyme

Marilyn Anderson

To a child's eye a lightning bug outshines the
brightest fixed star.

Oscar Penn Fitzgerald

Marilyn Kent

Snug as a bug in a rug .

Ben Franklin
Francis Gentleman

Mary Lee Dunn

All things bright and beautiful, all creatures great and small,
all things wise and wonderful, the Lord God made them all.

Cecil Francis Alexander

Steve LeLaurin, we thank you for your guidance and encouragement.

ISBN-13: 978-1502508584
ISBN-10: 1502508583

Marigold The Beetle Bugs Out

By
Marilyn Anderson
Marilyn Kent

Illustrated by
Mary Lee Dunn

In a morning glory field so very bright,
lived the beetle, Marigold

and all was just right.

Then the meadow was mowed
to be plowed and sowed.

"There goes my abode!" she said with a fright.

"I have dozens and dozens
of cousins and cousins."

So into the world she flew...

As Marigold landed on a truck of blue,
her tortoise shell began to change its hue.

She was excited and afraid,
but to the course she stayed

searching for a home that was brand new.

Flying south, Marigold did spy
acres and acres of farms passing by.

Red beetles on tomatoes,
green beans and potatoes.

She said "I think I'll give this place a try."

Lottie the Ladybug said,
"Please stay with me.
The farmers like us and let us be."

We eat the aphid pests
and other uninvited guests.

"Must go!" said Marigold,
"I only eat plants you see."

"I have dozens and dozens
of cousins and cousins."

So into the world she flew...

Traveling on a boat of blue,
she heard a rumor – is it true?

A beetle so large, as big as a barge,
So to the Amazon is where she flew.

Meeting Rio, a Titan beetle,
was quite an event.

"It must take a lot to daily fill your pot."

"Adults don't eat", Rio sighed with content.

Marigold was hungry and sad,
"These jungle plants taste
yucky bad. I'll get my treat,
since you do not eat,

and wait for the morning glories
to make me glad."

"I have dozens and dozens
of cousins and cousins."

So into the world she flew...

Marigold slept in a blanket of blue.

It covered the produce the farmer grew.

Onions, radishes and beets,
green peppers and other treats

at market the farmer hoped to get his due.

Juan was a Mexican Bean beetle party bug.
At markets, he would show his handsome mug.

Colors and noise,
food and rowdy boys.

"Too much for me,
so goodbye with a hug."

"I have dozens and dozens
of cousins and cousins."

So into the world she flew ...

Jetting into a sky of blue,
Marigold flew to a place that
was quite new.

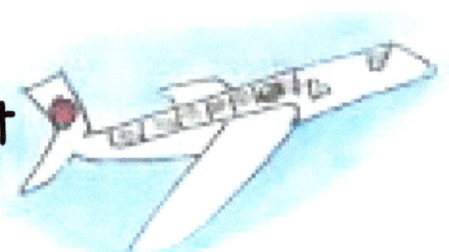

Hidden in a produce crate,

She was excited but had to wait

to meet her Japanese cousin crew.

Goro took Marigold to a rose feast
with thousands of beetles per plant, at least.

Pushing and shoving to get a bite,
"Eating like this, it just isn't right.
Nice meeting you, must go!
My time with you has ceased."

"I have dozens and dozens
of cousins and cousins."

So into the world she flew...

Land, air and sea of blue,
she sticks to each
transport like glue.

Australia bound,
beetles all around.

"Maybe I'll find a home that will do."

Aussie beetle called out, "G'Day."
"I'm a Tiger beetle and I like to play.
My mates call me Mick
and I am very, very quick.

Be careful or you might be my dinner today."

"Rooaarr"

"ooch!"

Ooooh!

"I have dozens and dozens
of cousins and cousins."

So into the world she flew...

Crossing deserts and savannahs
on a train of blue,
mountains, jungles, seas and swamplands too.

Marigold traveled far,
also by boat and car
stopping in a royal palace of golden hue.

Cleo was a Scarab beetle of royal birth.

Scarabs were copied on treasures
of great worth.

"Who are you and where is your poo?"
Cleo asked Marigold with little mirth.

"All Scarab beetles into tunnels roll dung
to provide the food that feeds our young.

Don't judge with haste, we use the waste
or else all would be piled to the highest rung."

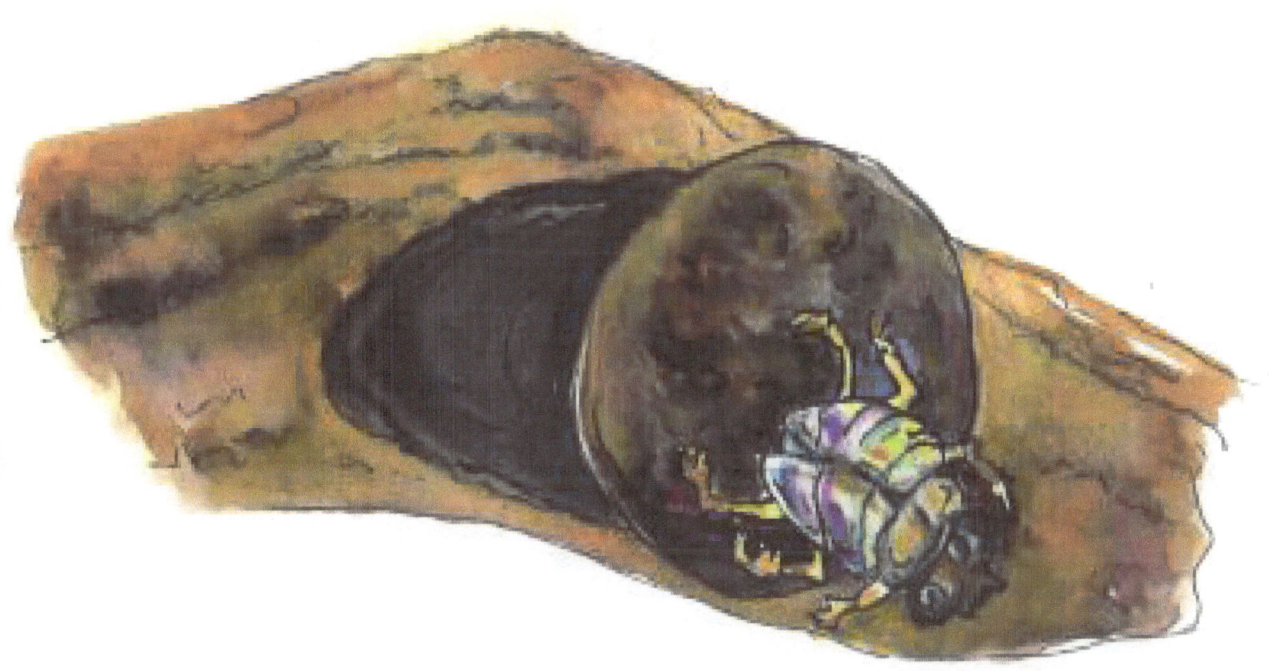

"I have dozens and dozens
of cousins and cousins."

So into the world she flew...

Heading north,
Marigold took her cue.
A beetle magic land to pursue.

Perhaps this is her new home, a
place to settle, no more to roam.

So off to Germany she flew...

"Guten Tag, mein little pretty one,
I'm Hans, a German Stag, who gets things
done, like finding the right spot
for all the beetles in the lot.
you are number four,
isn't this fun!"

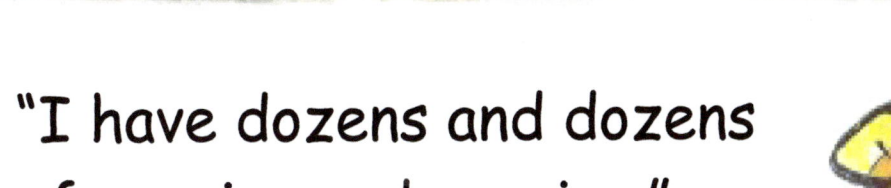

eins   zwei   drei   vier

"I have dozens and dozens
of cousins and cousins."

So into the world she flew...

Then Marigold boarded an ocean liner of blue,

She rode 'cross the Atlantic,
searching for something new.

The wind did blow,
the waves tossed to and fro;

"I'll find my home this time it's true."

As Marigold neared the South Carolina coast,
Beau, a Giant Diving beetle,
offered to be her host.

She told him her tale, trying hard not to wail.

Beau said, "This is the place I like most."

"A famous author named Edgar Allan Poe lived on this island a long time ago.

My dear, but you look like the goldbug in his book.

Come with me and see," wooed Beau.

Emma, a Firefly beetle,
lived in the field nearby.

From the bed of morning glories
she would arise at night and fly.

Her sparkle was bright,

blinking throughout
the night

lighting up the
nighttime sky.

Beau took Marigold to
Emma's morning glory field.
"My new home!" she shouted and squealed.
"There's nothing more I could wish or ask for
Beau, to your lovely charms I do yield."

"I have dozens and dozens
of cousins and cousins.
Into the world I need not go,
staying in my new home
just me and my Beau."

USA

Germany

MEXICO

EGYPT

JAPAN

Brasil

Australia

Antarctia

Start

Finish

Beetle Cousins Come In

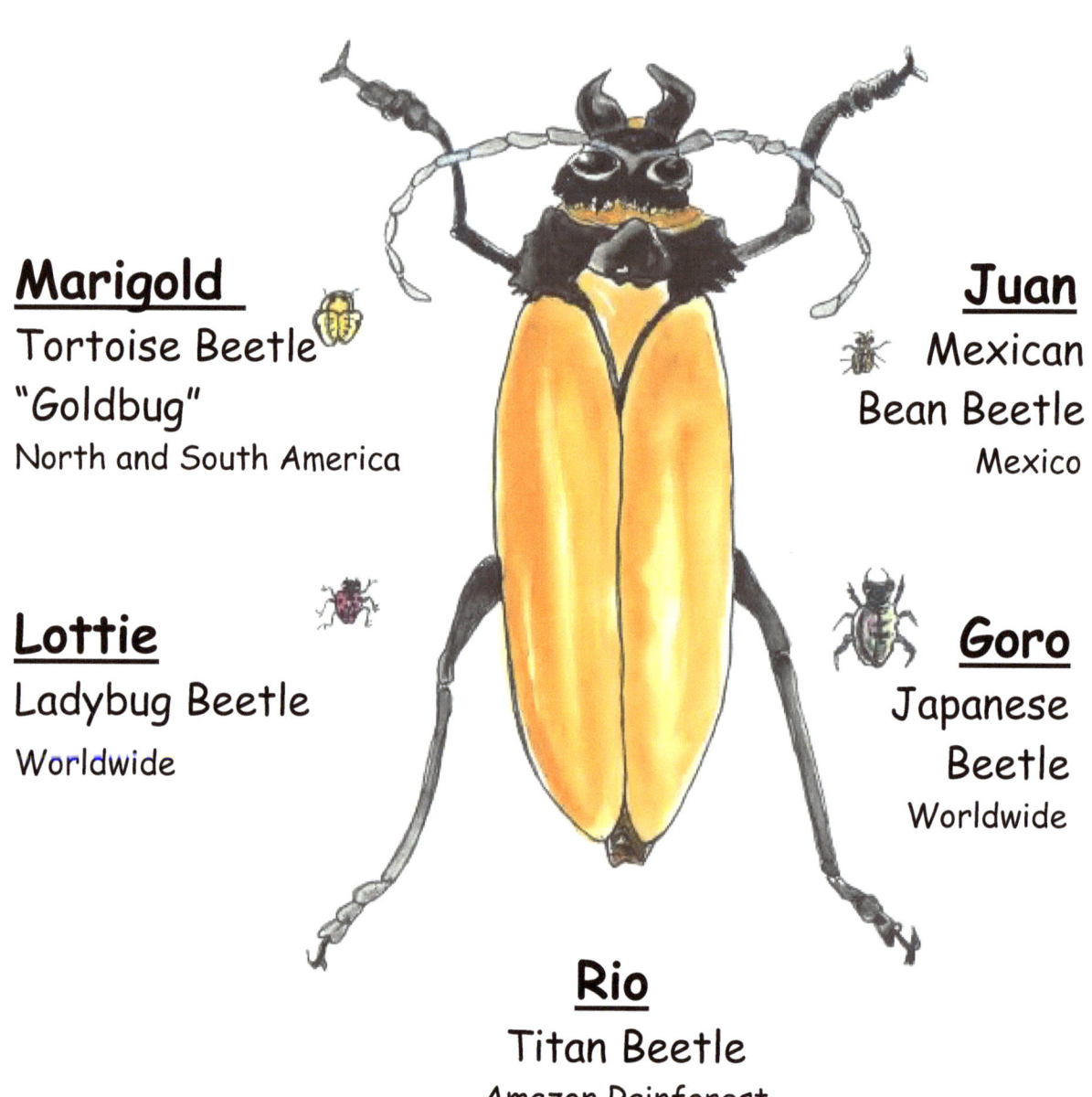

Marigold
Tortoise Beetle
"Goldbug"
North and South America

Juan
Mexican
Bean Beetle
Mexico

Lottie
Ladybug Beetle
Worldwide

Goro
Japanese
Beetle
Worldwide

Rio
Titan Beetle
Amazon Rainforest

Many Shapes, Sizes and Colors

Mick
Australian
Tiger Beetle

Australia

Beau
Giant
Diving
Beetle
North
America
Europe and Asia

Cleo
Scarab Beetle

Egypt and All
Continents
Except Antarctica

Emma
Firefly
Beetle
Worldwide

Hans
German Stag Beetle

Europe

Golden Tortoise Beetle   MARIGOLD

The golden tortoise beetle is in the leaf beetle family and lives in the Americas. Morning glories and sweet pea foliage are associated with the golden tortoise beetle. It is 5-7 mm in length and can be gold (goldbug) to orange and it is often metallic. The color changes as the seasons change. When threatened, the golden tortoise beetle will turn from shiny gold to a dull brown. The golden tortoise beetle lays approximately 20 flat white eggs on foliage.

Size 3-7 mm(.12-.25 inches)   Order is Coleoptera   Family is Chrysomelidae

Ladybug Beetle "Ladybirds"   LOTTIE

The ladybugs are small insects that are yellow, orange and scarlet with black wing spots, black legs, head and antennae. The ladybug can be found in the United States, the United Kingdom, Ireland, Canada, Australia, Sri Lanka, Pakistan, South Africa, New Zealand, India and Malta. They feed on aphids and scale insects that are garden pests. Some do eat plant leaves. The number of wing spots does not indicate age and vary from species to species.

Fun Fact! A lady bug can eat up to 5,000 aphids during a lifetime. The bright colors of a ladybug warn predators. That is, the red and black colors are not appetizing to predators. Ladybug larvae look like tiny alligators.

Size 1-10 mm (.05-.40 inches)   Order is Coleoptera   Family is Coccinellidae

Titan Beetle RIO

The titan beetle is the largest species of beetle in the world. Sometimes, it is mistaken for a cockroach. It is found in the Amazon rainforest in hot humid regions very near the equator. It is a member of the genus Titanus and belongs in the family of longhorn beetles.

Fun Fact! The titan is so large that it resembles the section of a vacuum cleaner hose.

Size 16.7 cm (6.5 inches)   Order is Coleoptera   Family is Cerambycidae

Mexican Bean Beetle   JUAN CARLOS GONZALES

The Mexican bean beetle is a plant feeding lady beetle found in the Southwest, Idaho, South Dakota and parts of the Eastern United States. It is believed that the first Mexican Bean beetle arrived in Colorado around 1853. Hay was delivered to the cavalry horses during the Mexican-American War in 1848 and the beetle traveled in the hay. It is known to damage seed pods and leaves. The shape of the Mexican bean beetle is convex and covered with tiny hairs called setae. The color of the beetle ranges from pale yellow to dull brown and it has eight small black spots on each elytra.

Size 6-7 mm (.25 inches)   Order is Coleoptera   Family is Coccinellidae

Japanese Beetle   GORO

The Japanese beetle is the most destructive of the Scarabaeidae family. It can be found throughout the world. The beetle was first discovered at a New Jersey nursery in 1916. It is believed that the larvae entered the United States in a crate of iris bulbs. The larvae of the Japanese beetle causes damage to the root systems in lawns, golf courses, parks and pastures. The adults feed on flowers, foliage and fruit of as many as 300 species of plants and trees. The beetles are most active on sunny days and cause severe damage when they eat in groups.

Size 6-13 mm (.25-.50 inches)   Order is Coleoptera   Family is Scarabaeidae

Australian Tiger Beetle MICK

The Australian tiger beetle is known for its running speed and predatory habits. The fastest can run at 5.6 miles per hour. When compared to a human runner, that would be 480 miles per hour! Most are iridescent blue or yellow. Tiger beetles have large bulging eyes, long legs, and curved jaws. The quick moving adults run down their prey and live along sea and lake shores, sand dunes or wooded paths.

Size 15-20 mm (.63-.75 inches)   Order is Coleoptera   Family is Carabidae

Dung or Scarab Beetle CLEO

The dung beetle is also known as the scarab beetle. The dung beetle primarily feeds on feces. They are usually brightly colored and can fly in a zigzag pattern in search of dung. They are known as tunnelers, rollers and dwellers and live in manure. The dung beetle can live in hardwood forests, farms, deserts as well as grassy open areas. The adult can burrow to lay eggs and store dung for food. The dung beetle can be found on all continents except Antarctica. It is especially good at controlling flies and harmful worms in cattle dung. The dung beetle also improves soil fertility.

Fun Fact! The dung beetle can roll a ball 50 times its weight. In fact, the male, Onthophagus Taurus can pull 1,141 times his own body weight! That is the equivalent of a person pulling six double decker buses full of people! Usually the male rolls the food. In some cases, the male and female roll the food together. When they find some soil that is soft, they bury the ball and can use it later for laying eggs inside and providing food for the offspring.

Size 11-22 mm (.43-.86 inches)   Order is Coleoptera   Family is Scarabaeidae

German Stag Beetle HANS

The German stag beetle lives in European countries and is part of a family best known for their large mandibles (jaws) that look like antlers of a stag. Male stag beetles wrestle each other like the male deer fight over females. The larvae feed on rotting wood and grow through three larval stages. In the final stage, the grubs can be as large as a human finger!

Size 12 cm (4.8 inches)   Order is Coleoptera   Family is Lucanidae

Giant Diving Beetle   BEAU

The giant diving beetle is part of a predatory family that dives in the water to prey on food larger than themselves such as frogs and toads, salamanders and small fish. The adults can tear their prey apart. This diving beetle can be found in Europe, North America and Asia living in ponds, lakes, streams and rivers. Most are brown or black with a light brown or red border. The long hind legs are hairy and used for swimming.

Size 1.5mm-40mm (.06-1.6 inches)   Order is Coleoptera   Family is Dytiscidae

Firefly/Lightning Bug   EMMA

The firefly is a winged beetle that produces a "cold light" from a chemical reaction in the lower abdomen. This phenomenon is called bioluminescence. It may be yellow green to pale red. Each species has a distinctive pattern of flashes and can be recognized by the number, interval and duration between them. The 136 species can be found in marshes or wet areas in North America and throughout the world when the larvae find food. Some species do not flash and can be found on branches of trees or on flowers. Those that flash at night rest during the day on leaves.

Size 4.5- 20 mm (.17-.8 inches)   Order is Coleoptera   Family is Lampyridae"lamp"

Fun Facts

There are 1,032,000 animal species.

750,000 of these species are insects!

40% of all insects are beetles. Estimates range from 350,000 to 400,000.

Coleoptera   (Coe-lee-OP-ter-rah) is the order.

Activities: Beetles can swim, bore, burrow and some species can fly!

Food: Beetles eat a variety of plants, fungi, and animal remains. Some are parasites during the larval stage.

Habitat: Beetles are found from the mountains to the beaches.

Body Structure: Some beetles have tough hard bodies and some are soft, shiny, brightly colored, and metallic. The head has mandibles (jaws) for chewing.

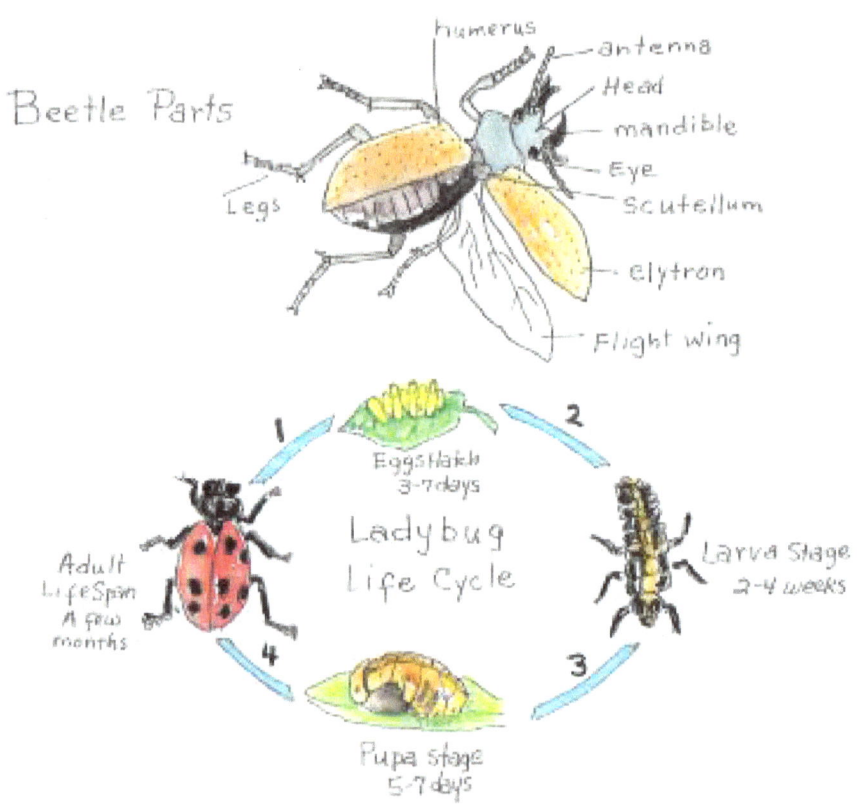

BEETLE WORD SEARCH

```
E B M E T A M O R P H O S I S
U I E M A I I A A P U P A E O
G O C U A S E R O R T B A C D
E L O T A E A A O O D N L C P
M U M O A L P E A O N N L R S
E M P N R B M X M E N A M S C
E I O O R I I E T T R S L U P
E N U R R D N N R V M S N N A
N E N P R N A T A T L R G U R
T S D E R A N E R M N S L H T
H C E X I M A D O C N N G S Y
O E Y A N T O N O T N G B Y L
R N E A E U G G E T L U D A E
A C S S I E S A O E B S L A M
X E E G N E O U N T N S C D B
```

FIND THE VERTICAL, HORIZONTAL, DIAGONAL AND BACKWARDS HIDDEN WORDS:

ABDOMEN, ADULT EGG, ANTENNAE, BIOLUMINESCENCE, COMPOUND EYES, ELYTRA, LARVA, MANDIBLES, METAMORPHOSIS, PRONOTUM, PUPA, SETAE AND THORAX.

Glossary

Abdomen-The most posterior portion of the insect body.

Adult-The mature insect/beetle. The last stage of metamorphosis.

Antennae-The sensory organs on the head. They detect food, sites to lay eggs and mates.

Bioluminescence-A chemical reaction that produces light in fireflies.

Compound eyes-In the adult, the eye is composed of dozens, hundreds, even thousands of lenses used to see images.

Egg-The first stage of metamorphosis.

Elytra-The hard wing case of the beetle. The wings are under the tough elytra.

Flight wings-The adult outgrowths of the beetle exoskeleton.

Humerus-The "shoulder" of the beetle.

Larva-Name of young insect.

Legs-Three pairs of legs

Mandibles-The jaws of the insect.

Metamorphosis-The four main stages of development i.e. egg, larva, pupa, and adult.

Pronotum-The back surface of the first part of the insect. It is the body part between the head and wing.

Pupa-The third stage of the life cycle.

Scutellum-The rear or posterior portion of the beetle thorax.

Setae-Tiny hairs.

Thorax-The middle section of the three parts of the insect body. Wings and legs are connected to the thorax.

www.ingramcontent.com/pod-product-compliance
Lightning Source LLC
Chambersburg PA
CBHW050404180526
45159CB00005B/2140